MW00965318

Ham Radio:

Easy Quick Start Ham Radio Reference

Table of Contents

Introduction

Ham Radio, also known by the name of Amateur Radio, is the hobby or craze that is overtaking America and some of the other countries by its popularity. Ham Radio basically includes the communication that is held with the help of radio equipment. This communication is usually done by the people who are licensed to do this communication.

People communicate through radio due to various reasons. Some of the people tend to do this solely for enjoyment purposes; they adopt it as their hobby. Some of the people do this for public service while there are also many people who self-train themselves through this mode of communication.

Ham Radio is commonly understood as a community of those people who communicate with other people through a set of radio equipment. The people who are part of this community are known as ham radio operators. The most interesting aspect of this activity is that it is done without the help of any Wi-Fi or internet connection.

If you are thinking that these people might be using cell phones to communicate with each other then you are totally wrong. All they use is a set of radio equipment; the details about this equipment will be discussed in the last chapter of this book. Through this book, you will also get to know the method to organize your own Ham Radio at your home.

Every person adopts the hobby of being a ham radio operator for different reason. The only thing common in all of these ham radio operators is that they totally love it. All of these ham radio operators have to get license from the government in order to pursue their hobby as a ham operator.

If you are thinking that by becoming a ham radio operator, you might be able to broadcast different programs to the public then you are wrong. Once you become a ham radio operator, you will only be able to communicate with different ham operators using the suitable codes and frequency. Just read this book and find out more about ham radio.

Chapter 1- Key Concepts of Ham Radio

Before moving on to the key concepts of ham radio, we must first understand what ham radio is. Ham Radio is an amateur method of radio communication that allows the ham radio operators to communicate with each other by using a specific coded language. Ham radio operators are required to get a certain license in order to practice as a ham radio operator. This license is only provided to these people if they pass the required exams. Ham operators also introduce themselves as "HAMS".

The communication of ham radio operators is not just any random communication but it is actually a legalized and controlled form of communication. All the services of ham radio are controlled by the International Telecommunication Union. The ITU controls and regulates the ham radio services with the help of International Amateur Radio Union.

One thing that you need to keep in mind about the ham radio operators is that they are also known as amateur radio operators for a reason. These ham radio operators are not allowed to earn money through their activities on ham radio. Ham radio is just a hobby or a mode to offer public services.

All the licensed ham radio operators just use ham radio for the purpose of enjoyment or learning. A ham radio operator tends to communicate through Morse code or coded language. At first the communication will seem very complicated but once you learn the initials, everything starts sounding fun.

Some people also become CB radio operators. CB radio operators are non-licensed radio operators. CB radio operators in comparison to the CB radio operators don't get to enjoy any of the privileges attached with the license of ham radio operators.

Key Concepts Cleared

When we come to the key concepts of ham radio then there are some basic things that you must know before opting for the hobby of ham radio.

Who can become a ham radio operator?

Well the simple answer is that anyone can become a ham radio operator. It doesn't matter whether you are a doctor, a constructor, a house wife, a librarian or a student. People belonging to different fields of life actively enjoy the hobby of ham radio.

If you want to become a ham radio operator then it doesn't matter whether you are rich or poor, young or old; anybody can become a ham radio operator. There is not a requirement of any specific nationality or any specific religion for adopting the hobby of ham radio operator.

Not many years ago, these ham radio operators tend to use the brass telegraph key for developing a communication with other ham radio operators. But these days most of the ham radio operators have turned towards Morse code for communication.

Many ham radio operators also used hand-held radios in order to communicate with other ham radio operators. While nowadays ham radio operators are communicating through the computerized messaging. This type of messaging is conveyed with the help of satellite. But in every case, you are going to see one common thing and that is a radio. Every ham radio operator communicated with the help of a common radio.

Benefits attached to Ham Radio

After taking one look at the benefits of ham radio, you are definitely going to conclude that ham radio, without any doubt, is the most powerful means of communication. Ham radio can easily pass on your messages to the world where all other types of communication methods have failed.

These days, people think that the most reliable modes of communication are cell phones or internet. But even these things fail to pass on your message when the signals fall down or get blocked due to any reason.

When the incident of 9/11 hit America, internet signals got disrupted due to which most of the United States Federal agencies were not able to communicate with each other. In such a condition, ham radio was the only shelter that provided solace to these agencies.

The fact is that with ham radio, you don't need to get a proper setup or a studio in order to communicate with the other hams. Ham radio messaging can easily be done from any corner of this world. Whether you are stuck in a jungle, on a mountain, in flood or lost a way while riding in your car; all it takes is a radio to communicate with the world.

How can you communicate through Ham Radio?

The communication through ham radio is fairly simple. Most of the people either communicate using their voice with the help of a microphone while some of them prefer to communicate through

Morse code. If a person wants to send images through the ham radio then first he would have to interface his radio with his computer. Many people also enjoy the privilege of communicating with astronauts through ham radio. Here is a list of all those methods that are commonly used by the ham radio operators for communicating with other hand radio operators:

- Communicating through voice

- Communicating through Morse code

- Packet

- Communicating through television

- PSK

- Ritty

Reason behind getting a license for ham radio

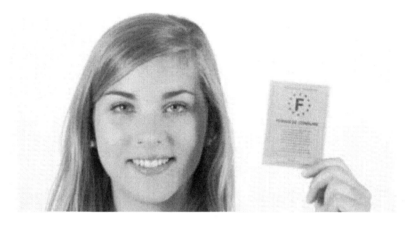

Most of the people get confused when it comes to the requirement of obtaining a license in order to be a ham radio operator. This requirement of getting a license was basically imposed by FCC or Federal Communications Commission. FCC felt that there was a need for a team of radio operators that could provide emergency communication services in a situation where all the other modes of communication fail.

Why are ham radio operators called by the name of "Hams"?

Although nobody knows the specific reason behind the word "Hams" used by the ham radio operators but one could relate it to the older times when the concept of ham radio was still new.

At that time, the signals passed on by the ham radio operators were extremely powerful that they even disrupted the signaled communication of commercial operators. Due to this disruption, the commercial operators got frustrated and started using the term "Hams" for the ham radio operators. This trend became more common and now present day ham radio operators proudly introduce themselves as "Hams".

Understanding Amateur Radio Bands

Amateur radio bands are the frequencies that are used by the hams to communicate with each other. If you are confused about what the frequencies are then let's understand this in a practical way. If you have an old radio then just take a look at it. You will see a sort of meter on your old radio with number markings on it.

These markings usually start from 535 kilohertz and end up at 1605 Kilohertz. This meter or number markings is understood as a radio band. Whether it's commercial radio operator, a military operator or a ham radio operator, all of them use the frequencies mentioned on the radio bands for the purpose of communicating. But not all of them use the same bands or frequencies.

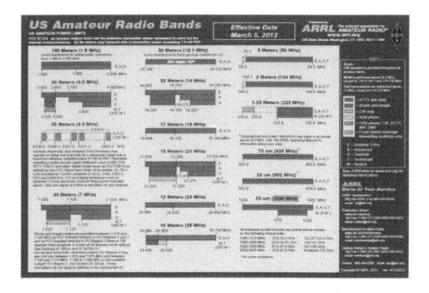

Various ranges of frequencies can be found on the radio band. These frequencies are usually recognized as "vlf" i.e. the lowest possible frequencies and "ehf" i.e. frequencies at extremely high range.

When it comes to the frequencies that are used by the ham radio operators, they are usually permitted to use 26 bands through which they enjoy communicating across their town, city or even into the space. The frequencies that are used by the ham radio operators usually start from 1.8 megahertz to 275 gigahertz.

Is it an expensive hobby?

Although it might seem like that but the answer is no, it's not an expensive hobby. You will need to spend 200 dollars at the maximum while receiving the training and settling up your gear for this hobby.

The initial training and the radio set-up might require a little bit of investment but after that you won't feel the need to spend money on your learning as most of the people tend to self-train themselves after they have started with ham radio.

Chapter 2 – Activities and Practices of Ham Radio

So what is it that all of these ham radio operators do? There are many people who are suffering from different misconceptions regarding ham radio and the sort of activities that are performed by these ham radio operators.

Here is a list of all those activities that are performed by the ham radio operators:

* Hams or Ham operators get to enjoy endless communication opportunities to any part of the world. If you are a ham radio operator then you can easily communicate with any other ham radio operator residing in any part of the world. These hams use high frequency radios for the purpose of communicating with the other hams.

- You can communicate with other ham operators in your town and get to make new friends through ham radio. Communicating within your town is very easy; all it takes is a VHF transceiver while some of the people also use UHF transceiver. These transceivers are very small and easy-to-carry.

- Hams also get to communicate on the low power also known as QRP. Although it might seem extremely challenging but there are many hams who seem to love communicating on such low power. Hams tend to practice this low power communication on the high frequency bands.

- Packet radio is another thing that is enjoyed by most of the hams. It is a digital form of communication where the hams tend to send data from one radio station to another radio station. This type of communication is done with the help of a terminal node controller.

- Most of the people have this misconception that in order to become a ham radio operator, one must know how to communicate in Morse code. Well not necessarily. Although there are many hams who prefer to communicate in Morse code but there are also some of them who don't feel the need for this coded language.

- Hams also get to enjoy the communication through television.

- If you want to send images or any of your snaps to another ham radio operator then just use "Slow-scan television" technique. Ham radio operators usually adopt this technique when they have to transfer the images without spending a dime.

- Most of the hams also hold different contests or competitions through their ham radio transmission. During these contests, they tend to compete with other hams on the basis of their ham radio skills.

- Hams also tend to provide emergency contact services in the event where all the other modes of communication have failed due to any natural disaster. During the hurricane Katrina, hams used their ham radios in order to call for help.

- Hams can also perform communication via satellites. You will also be amazed to know that ham radio operators have their own satellites. Communicating through these satellites is easier than communicating through other satellites.

- Hams also play the role of free-of-cost messengers where they usually convey messages from people to people living in different parts of world.

- After becoming a ham radio operator, you can even order pizza with your ham radio portable set. People also used to make calls to hospitals, doctors or police stations using the ham radios.

Apart from all the above mentioned activities and practices, hams also get to enjoy round-table discussions on regular basis. These discussions include latest happenings in the field of ham radio or issues of common interest.

Chapter 3 – Licensing Requirements for Ham Radio

If you are interested in becoming a ham radio operator then the very first step that you need to take is to get a license for ham radio. While activating as a ham radio operator, you must know and follow the rules otherwise, you will end up having your license cancelled. For ham operators residing in USA, there are three different types of ham radio licenses offered by the FCC.

- The first type of license is known as Technician license. If you are a newbie then you ought to get the technician license. In order to attain this license, you need to pass one exam. This exam will comprise of 35 questions. This exam usually tests the people for their understanding of theoretical knowledge of radio, regulations that need to be followed as a ham radio operator and the operational activities that are performed by a ham radio operator. Once you get this license, you will be allowed to develop a communication on the frequency of 30 megahertz and above. Most of hams possessing this license are able to communicate with the local hams residing within the boundaries of North America.

- Next comes the General license. In order to get this license, you must first possess a technician license. You also have to pass a exam which is completely a written one. This exam will also test you for 35 questions. Once you pass this exam, you get to enjoy the privilege of having a world-wide ham radio communication.

- The third type of license offered by the FCC is known as the Extra-class license. For this license, you are required to get all the other licenses before it. In order to attain this license, you will have to pass through an exam in which you will be required to answer 50 MCQs. You won't be required to learn Morse code for this exam. On the other hand, you will be required to clear your concepts about advanced ham radio regulations, advanced level electronics, specialized and advanced operating activities, understanding the radio designs.

How to prepare for the licensing exams?

Bear this in mind that all the ham radio licenses are issued by FCC. You will be amazed to know that most or almost all of these tests are not conducted by the FCC. It is the expert hams who voluntarily conduct these tests and include normal people into their hams community. So you get to take your ham radio exam under the supervision of an expert ham radio operator. Once you are done with the exam, your supervisor will submit your paper and other required documents to the FCC. If you at least get a "D" then your license will be mailed to you.

After learning about this fact, you might be feeling a bit confused about how can you take this test or how can you prepare for this test. Well the entire procedure is quite simple as there are many ways through which you can be prepare for a ham radio licensing test.

The first method is that you can join a ham radio club. Joining a club is sort of an organized form of taking this test. All you need to do is to find out about the local clubs that are conducting their classes for these types of tests. You attend the classes and at the end you have to take the test. Once you pass, you are awarded with the license.

Second method for taking a license exam is to get prepared under the guidance of a trainer. These trainers (also known as Elmers) know all about the exams and will guide you at every step to help you get the license.

Last but not the least, if you are unable to find the first two methods suitable then you can also do this through self study. Although most of the hams preferably use the first two methods, but the third method is also used by lots of people who have busy work-schedule.

Chapter 4 – Understand Morse Code for Ham Radio

Created by Samuel F.B. Morse, Morse code is the most preferred mode of communication by the ham radio operators. Other than communicating through the ham radio, Morse code is also used for various purposes. It can be used for communicating a message in routine life through a telegraph or it can be used for conveying an emergency message. A Morse code message can also be signaled with the help of a mirror's flash or the light of a torch.

Although Morse code is not compulsory for ham radio but if you still want to learn Morse code then it can easily be done by following these steps.

* Morse code mostly comprises of dit's and dah's used in different patterns. When you have to convey any alphabet with a short beep then it will be

understood as dit, while the alphabets that are conveyed with longer beeps are regarded as dah. While communicating an alphabet, a small pause is observed but if you are about to communicate the next word then you will take a long pause.

- The first ever method that you can use to learn the Morse code is to listen to the recordings of Morse code. It's best if you can get your hands on slow recordings as they will help you understand a lot easily. You can listen to the Morse code recordings by downloading Morse code software. Such softwares are usually free or available at very cheap prices.

- Try to concentrate on the sounds of the alphabets or words while you listen to the Morse code recordings.

- You can also memorize the Morse code with the help of a chart depicting the Morse code alphabets. There are two types of charts that people use to learn the Morse code. The first one is known as the simple or basic chart through which you can learn the alphabets of Morse code. The second is the advanced form of the Morse code chart which will also help you learn various abbreviations or signs used in the Morse code.

- If you find it comfortable then you can also learn these signs and alphabets by writing them down. Most of the hams don't find it necessary to follow this writing step but it will help you learn quickly if you do so.

- Just focus on sounding out the words. Go for easiest alphabets at the start and make a start with smallest of the words. First listen, then write it down and then try to speak it out. Gradually try to eliminate the writing step from your procedure because at the end you will just have to pronounce the sound of it.

At the end all you will have to do is practice, practice and practice.

Chapter 5 – Tips to Assemble Your Own Station

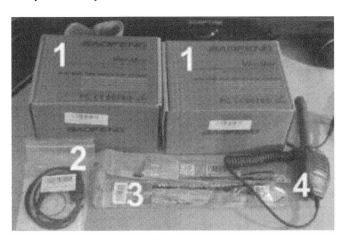

If you want to assemble your own ham radio station then it is not that much difficult. In the image mentioned above, are the radio boxes pointed out as number 1. These boxes also contain the plug-in chargers, drop-in charger, and a flexible small-sized antenna. These radio boxes also contain an instrument manual which is not easy-to-understand unless a person is highly trained in electronics.

The small-sized bag labeled as number 2 basically contains a small-sized CD. Using this CD is really simple. This CD is just inserted into the concerned part of the compatible drive and then the software is installed.

The items marked at number 3 in the above image, are the antennas that will come quite handy while creating your own ham studio. These are large-sized antennas; large antennas are used when you want to communicate to the ham radio operators residing in the other parts of the world.

Some people think that large antennas are compulsory for around-the-world but mini-antennas are not important, this is false. There are many situations where mini-antennas become very useful instead of large antennas.

If you want to communicate with a person residing in the same building as you then min-antenna will be extremely beneficial for you. But if you want to communicate with a person living across the town or any nearby town then large-sized antennas are best.

The item marked as number 4 in the image is a microphone. Such microphones are usually used by the people when they are on the hike into the forests or onto the mountains. Those hams that use this technology while wandering outside don't have to reveal the radio to other people.

The radios are usually hidden under the belts while most of the people hang their microphones near their collar or on their shoulders. The noise level is extremely high when we are hiking on a mountain or having a trip into the forest. In such an instance, getting the chance to use the microphone near your ear seems like a great blessing. If you think that using a microphone seems like a noisy activity then you can also use Morse code to convey your message.

Ok now let's enter the phase of putting this gear together so you can set up your own station. This phase is very easy and simple, all you need to do is open the radio boxes and remove the radio from these boxes. Next you must take the main power wire and plug its switch into the charger. If the wire has been properly plugged into the charger then the charger lights will start blinking green.

Now you will have to place the radio into the charger. You will see an inbuilt space into the charger where the radio needs to be placed. If you look closely under the radio, you will some circuits that will come into the contact with circuits built in the charger once you place the radio in it.

If the radio is properly placed into the charger then the light on the charger will display red color. Now you should let it charge and in the meanwhile turn your attention towards the programming.

While starting at the programming, you will come across a specifications manual for you radio set. The first step is to make an analysis of your location and the ratio of radio channel activity in that area. For instance, if you ponder over your location while residing near a huge city then you will come to know that your area has extremely high ratio of radio channel.

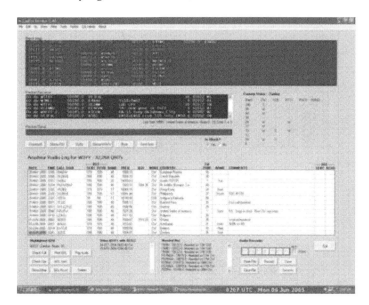

First of all, you need to get information on the radio band frequency that might be suitable for your radio in such a busy area. Usually you have to make your choice from the frequency of 136 megahertz to 174 megahertz. Some of the people won't find it necessary pay this much attention to this step. If you won't strive at this step then your radio transmission will be open to various threats.

Next you need to set your radio channels for going on-air. First you need to decide that how many channels do you need and also note down the purpose of those channels. After you have cleared these basic points, then you can mark the numbers of these channels. Hams use the first channel for the purposes of communicating purposes.

This channel requires you to have a high level antenna. The second channel that most of the hams use is known as NOAA, it is a weather channel and almost every ham radio operator has access to this channel. The third channel is known as Tactical Coms. This channel is used for emergency purposes in order to get into the contact with law enforcement agencies.

Fourth channel is the news channel while the fifth channel is for entertainment purposes. The fifth channel is basically a FM commercial channel used for hearing songs or learning about latest new happenings around the world. The hams are allowed to set up to 128 channels. Learning about the 128 channels might make you think that you will have a lot of options but when you start allotting channels to these numbers, you will be amazed at how less do they seem.

Some of the things that you need to keep in mind is that radio can only be used when it has been charged properly. You also have to program the radio in a proper way. In order to program the radio, you first need to sort out the required frequencies and types of channels that you want for your radio. You have to organize the channels in a group but first you need to sort out the grouping sequence for your channels. While programming a radio, don't forget to keep the frequency range limits in your mind.

Once you have finished with the programming which involves finalizing the channels and up loading them on the radio, you will be able to start practicing as a ham radio operator. Programming is a very lengthy and time-taking task so it's much better for you to get an already-programmed radio. You can easily get one from the hams club in your area or you can seek help from any of the local ham radio operators.

Conclusion

Thank you for downloading this book. This book is a valuable asset for all those people who have keen interest in the hobby of ham radio. After reading this book, you might not only be able to get an insight into the fundamentals of ham radio but you will also be able to understand how to become a ham radio operator. This book not only clears your concepts regarding the ham radio activities but it also provides you with the clear guidelines on how to get a license and set up your own personal station for ham radio. If you are seeking for a productive hobby then becoming a ham radio operator is the cheapest and most productive hobby on earth.

After reading through the first four chapters of this book, you might be able to understand how interesting this hobby is. Most of the people think that assembling a ham radio station at your premises is a complicated task that's why the fifth chapter of this book removes all such complications.

Ham radio operators don't only tend to have fun with their ham radio activities; they also feel a sense of responsibility towards the general public. After becoming a ham radio operator, you are definitely going to feel a sense of pride and responsibility. In the past, many hams have played an important role in serving the humanity. After reading this book, if you do think of becoming a ham radio operator then don't forget that you will be a part of the community that prides itself on helping those in need. So just read this book, become a ham radio operator and serve the humanity.

FREE Bonus Reminder

If you have not grabbed it yet, please go ahead and download your special bonus report *"DIY Projects. 13 Useful & Easy To Make DIY Projects To Save Money & Improve Your Home!"*

Simply Click the Button Below

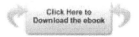

OR **Go to This Page**

http://diyhomecraft.com/free

BONUS #2: More Free & Discounted Books

Do you want to receive more Free & Discounted Books?

We have a mailing list where we send out our new Books when they go free or with a discount on Kindle. Click on the link below to sign up for Free & Discount Book Promotions.

=> Sign Up for Free & Discount Book Promotions <=

OR Go to this URL

http://zbit.ly/1WBb1Ek

Manufactured by Amazon.ca
Bolton, ON

28079696R00022